Books should be returned on or before the
last date stamped below

18. JUN 03.

03. 03. 04.

27.

19. SEP 05.

18. FEB 06.

2 8 MAY 2011
2 2 MAY 2013

1 2 AUG 2013

19. JUN 04.

2 8 SEP 2006

1 8 JUN 2014

31. JUL 04.

H O

1 8 AUG 2014

1 0 FEB 2007

13. SEP 04

1 6 OCT 2007

2 0 JUN 2017

09. OCT 04

1 0 NOV 2008

2 1 AUG 2009

16. MAY 05.

1 2 DEC 2009

Editor	Karen Barker
Language Consultant	Betty Root
Natural History Consultant	Dr Gerald Legg

Carolyn Scrace is a graduate of Brighton College of Art, specialising in design and illustration. She has worked in animation, advertising and children's fiction and non-fiction. She is a major contributor to the popular *Worldwise* series and *The X-ray Picture Book* series, particularly **Amazing Animals**, **Your Body** and **Dinosaurs**.

Betty Root was the Director of the Reading and Language Information Centre at the University of Reading for over twenty years. She has worked on numerous children's books, both fiction and non-fiction, and has also held the position of Smarties Book Award Judge.

Dr Gerald Legg holds a doctorate in zoology from Manchester University. His current position is biologist at the Booth Museum of Natural History in Brighton.

David Salariya was born in Dundee, Scotland, where he studied illustration and printmaking, concentrating on book design in his post-graduate year. He has designed and created many new series of children's books for publishers in the U.K. and overseas.

Printed in Belgium.

An SBC Book conceived, edited and designed by
The Salariya Book Company
25 Marlborough Place, Brighton BN1 1UB

©The Salariya Book Company Ltd MCMXCVIII

A CIP catalogue record for this book is available from the British Library

ISBN 0 7496 5031 1

This edition 2003
First published in Great Britain in 1999 by
Franklin Watts
96 Leonard Street
London
EC2A 4XD

Franklin Watts Australia
45-51 Huntley Street
Alexandria
NSW 2015

lifecycles

The Journey of a Butterfly

Written and Illustrated by Carolyn Scrace

Created and Designed by David Salariya

W
FRANKLIN WATTS
LONDON•SYDNEY

Butterflies are insects with long bodies
and large, brightly coloured wings.
In the autumn, monarch butterflies
fly from the cold north of America
to the warm south (see map, page 26).
There they spend the winter
half-asleep in the trees.
In the spring, they fly back north.

Their journey is called a *migration*.
In this book you can follow the
amazing migration of
the monarch butterfly.

During the summer, monarch butterflies live in Canada and the north of America.

They spend all summer feeding.

Antennae

Head

Wings

Monarch butterflies fly
over 3,000 kilometres
across America
during their migration.

In the autumn it gets cold.
There is less food for
butterflies to eat.

The butterflies gather in groups
to start the long
migration south.

The wings of a monarch butterfly are big and strong. They help it to fly quickly and for long distances.

The groups get larger as more and more butterflies join them.

At night, thousands of butterflies land on trees to rest.

Each morning
they continue
flying south.

In early winter
the butterflies reach
the warm south.
In their thousands,
they settle on the trunks
and branches of certain trees.
All through the winter
the butterflies stay there, half asleep.
This is called semi-hibernation.

The butterflies
use the same trees
every year.

In the spring,
the butterflies wake up.
They are ready
to start the long flight
back north.

For this return
journey, the
butterflies fly
on their own and
not in large groups.

During the journey north, male and female butterflies mate.

The females stop to lay their eggs along the way. Then the adult butterflies die.

The tiny eggs from the female butterfly stick to the leaves where they are laid.

Egg

Inside each egg,
a caterpillar grows.

Caterpillars hatch
from the eggs
and start to feed.
They eat
the leaves
of milkweed plants.

As the caterpillars get fatter
they grow new, larger skins.

Milkweed is poisonous
to other animals.
Eating milkweed
makes the caterpillars
poisonous too.
This means
other animals
do not eat them.

When the caterpillar is fully grown it hangs down from a leaf. After a few hours the caterpillar forms a hard case around its body.

Pupa case

It is now called a pupa.

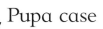

Pupa case

Inside the pupa case
the caterpillar changes
into a butterfly.
When it is ready,
the case splits open
and the butterfly
comes out.

The monarch butterflies that reach Canada and the north of America are the children of the butterflies that left there in the autumn.

In the summer, the monarch butterflies that have hatched from the eggs reach Canada and the north of America. The migration is over.

The butterflies live in the north
all summer. In the autumn,
it will be their turn to fly south.
The migration will start once again.

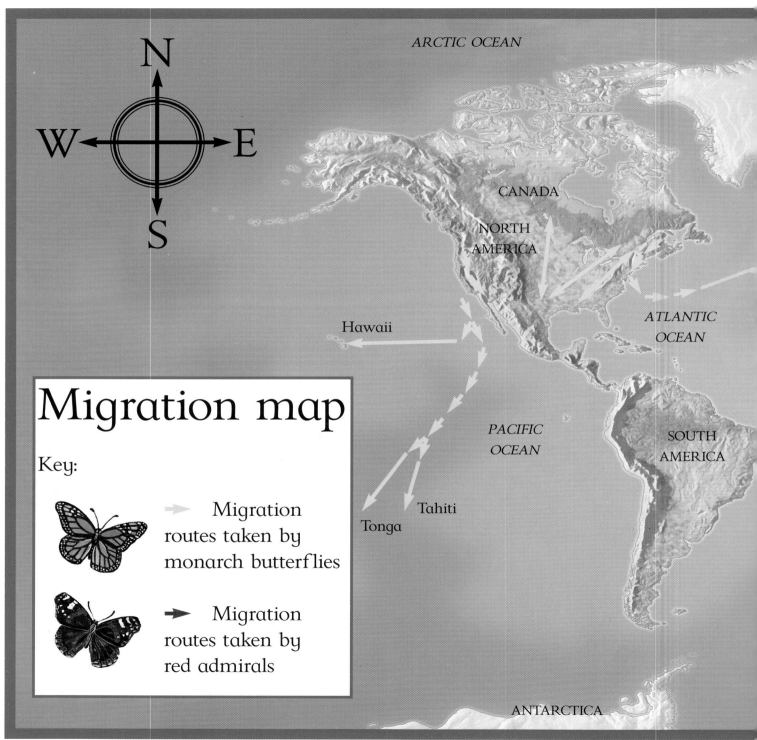

ARCTIC OCEAN

CANADA

NORTH
AMERICA

Hawaii

ATLANTIC
OCEAN

PACIFIC
OCEAN

SOUTH
AMERICA

Tahiti

Tonga

ANTARCTICA

Migration map

Key:

Migration routes taken by monarch butterflies

Migration routes taken by red admirals

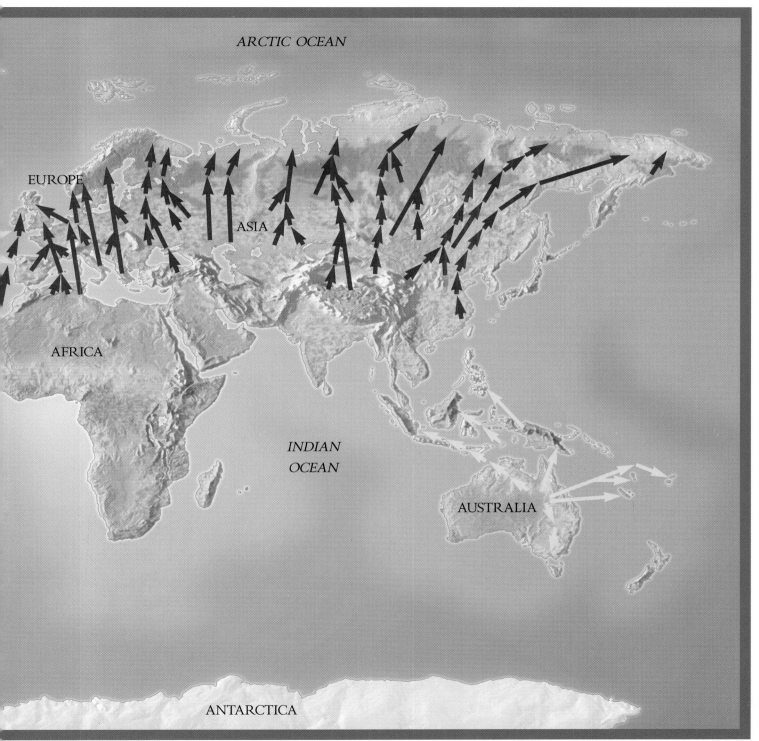

ARCTIC OCEAN

EUROPE

ASIA

AFRICA

INDIAN
OCEAN

AUSTRALIA

ANTARCTICA

Butterfly words

Antennae
The feelers on the head of an insect. They are used for smelling.

Caterpillar
The stage in the life of a butterfly between hatching from an egg and forming a pupa.

Flight
A journey made by flying through the air.

Hatch
When the caterpillar comes out of its egg.

Insect
An animal with six legs and two antennae. The body of an insect is divided into three parts and is covered with a hard skin.

Mating
When the male (father) and female (mother) join together to make a baby.

Poisonous
Something that makes most animals or people very ill if they eat it.

Pupa
The stage in the life of a butterfly between being a caterpillar and becoming an adult butterfly.

Semi-hibernation
When an animal is half-asleep all through winter.

Index